ERECTILE DYSFUNCTION GUIDE

A Comprehensive Guide to Managing and Overcoming Erectile Dysfunction for Better Health and Well-Being

Dr. Prieto O. Esteban

PUBLISHED BY:
Dr. Prieto O. Esteban

580 St. John Street, Somerset West, Western Cape. South Africa.

Title | Erectile Dysfunction Guide
Author | Dr. Prieto O. Esteban
First Print | 2024

Self-Publishing Titans
www.selfpublishingtitans.com
Made by human

Table of Contents

Chapter 1: Introduction

Definition of Erectile Dysfunction (ED)

Erectile Dysfunction (ED) is the inability to achieve or maintain an erection sufficient for satisfactory sexual performance. This condition can affect men of all ages but is increasingly prevalent with advancing age. It can result from a variety of physical, psychological, and lifestyle factors, including cardiovascular diseases, diabetes, hormonal imbalances, stress, anxiety, and substance abuse.

Importance of Understanding ED

Understanding ED is crucial for several reasons:

1. Health Indicator: ED can be an early warning sign of more serious health issues such as cardiovascular disease or diabetes. Addressing ED can lead to the diagnosis and treatment of these underlying conditions.

2. Quality of Life: ED can significantly impact a man's quality of life, affecting self-esteem, relationships, and mental health. Recognizing and addressing ED can improve overall well-being.

3. Treatment and Management: Knowledge about ED enables individuals and healthcare providers to explore and implement effective treatment options. This can include lifestyle changes, medication, therapy, or other medical interventions.

4.   Reducing   Stigma:
Increasing   awareness   and
understanding of ED helps
reduce   the   stigma   and
encourages more men to seek
help   and   discuss   their
condition   openly   with
healthcare professionals.

Scope of the Guide

This guide aims to provide a
comprehensive   overview   of
Erectile Dysfunction, covering
the following key areas:

1. Causes of ED: Detailed
examination of the physical,

psychological, and lifestyle factors contributing to ED.

2. Diagnosis and Assessment: Information on how ED is diagnosed, including medical history, physical examinations, and diagnostic tests.

3. Treatment Options: Overview of available treatments, including medications, lifestyle changes, psychological support, and surgical options.

4. Prevention and Lifestyle Changes: Tips and strategies

for preventing ED and improving overall sexual health through lifestyle modifications.

5. Support and Resources: Guidance on where to seek help, including professional medical advice and support groups.

By providing this information, the guide aims to empower individuals with the knowledge necessary to understand, manage, and treat Erectile Dysfunction effectively.

Chapter 2: Understanding Erectile Dysfunction

Anatomy and Physiology of an Erection

An erection is a complex process involving psychological, neurological, vascular, and hormonal factors. The key components include:

1. Nervous System: Sexual arousal starts in the brain, where neural signals are sent through the spinal cord to the pelvic region. The

parasympathetic nervous system triggers the release of nitric oxide, a neurotransmitter that relaxes the smooth muscles of the penile arteries.

2. Vascular System: The relaxation of smooth muscles allows blood to flow into the corpora cavernosa, two cylindrical chambers within the penis. The blood fills these chambers, causing the penis to expand and become rigid.

3. Hormones: Testosterone plays a vital role in sexual desire and erectile function.

Adequate levels of this hormone are necessary for normal erectile response.

4.   Muscles: The ischiocavernosus and bulbospongiosus muscles help maintain the erection by compressing the veins that drain blood from the penis, keeping it engorged.

Mechanisms of Erectile Function

The process of achieving and maintaining an erection involves several steps:

1.   Initiation: Sexual stimulation   (physical   or mental)   triggers   nerve impulses in the brain.

2. Signal Transmission: These impulses travel through the spinal cord and pelvic nerves to the penis.

3. Vasodilation: Nitric oxide is released, causing the smooth muscles of the penile arteries to relax and dilate.

4.   Blood   Flow: Increased blood flow fills the corpora cavernosa, leading to penile engorgement and erection.

5. Maintenance: Contraction of penile muscles compresses the veins, reducing blood outflow and maintaining the erection.

6. Detumescence: After ejaculation or cessation of arousal, the smooth muscles contract, blood flow decreases, and the erection subsides.

Types of Erectile Dysfunction

Erectile Dysfunction can be classified into three main

types based on its causes and characteristics:

Primary ED

Primary ED refers to the condition where a man has never been able to achieve or sustain an erection sufficient for sexual intercourse. This form of ED is rare and usually has psychological or congenital origins. Causes may include:

- Severe psychological conditions such as anxiety or depression.

- Congenital anomalies or developmental issues with the penile structure.

- Hormonal imbalances present from a young age.

Secondary ED

Secondary ED is the most common type and occurs in men who previously had normal erectile function but now experience difficulties. Causes of secondary ED can be both physical and psychological, including:

- Physical Causes: Cardiovascular diseases,

diabetes, obesity, hypertension, high cholesterol, hormonal disorders, and side effects of certain medications.

- Psychological Causes: Stress, anxiety, depression, relationship problems, and performance anxiety.

Situational ED

Situational ED occurs only in specific situations or with particular partners. It is often linked to psychological factors and may include:

- Performance anxiety or fear of failure.

- Relationship issues or lack of attraction to a particular partner.

- Stress or fatigue.

- Substance use (alcohol, drugs).

Understanding the type of ED is crucial for determining the appropriate treatment and management strategies, as the underlying causes and effective interventions can vary significantly.

Chapter 3: Causes of Erectile Dysfunction

Physical Causes

Erectile Dysfunction can often be traced back to physical health issues that interfere with the normal blood flow, nerve function, or hormone levels necessary for an erection.

Cardiovascular Diseases

Cardiovascular diseases such as atherosclerosis (hardening of the arteries), high blood pressure, and high cholesterol can impair blood flow to the penis, making it difficult to achieve or maintain an erection. The health of the cardiovascular system is crucial for erectile function.

Diabetes

Diabetes can damage blood vessels and nerves that control erection. Men with diabetes are at a higher risk of developing ED, particularly if their blood sugar levels are not well managed. Diabetes can also lead to heart disease, which further increases the risk of ED.

Obesity

Obesity is associated with a range of health issues that can contribute to ED, including cardiovascular disease, diabetes, and hormonal imbalances. Excess weight can also reduce testosterone levels, which are critical for sexual function.

Hormonal Imbalances

Hormonal imbalances, particularly low levels of testosterone, can affect sexual desire and erectile function. Conditions such as hypogonadism, thyroid disorders, and hyperprolactinemia can contribute to ED.

Neurological Disorders

Neurological disorders such as multiple sclerosis, Parkinson's disease, spinal cord injuries, and stroke can disrupt the nerve signals necessary for an erection. Damage to the nerves in the pelvic region can also result in ED.

Chronic Kidney Disease

Chronic kidney disease can lead to hormonal changes,

nerve damage, and reduced blood flow, all of which can contribute to ED. Additionally, the fatigue and general malaise associated with kidney disease can reduce sexual desire.

Peyronie's Disease

Peyronie's disease is a condition in which fibrous scar tissue develops inside the penis, causing curved, painful erections. This can make sexual intercourse difficult and may lead to ED.

Psychological Causes

Psychological factors can significantly impact erectile function. Emotional and mental health plays a crucial role in sexual performance.

Stress

Chronic stress can interfere with sexual arousal and the ability to achieve an erection. Stress affects the body's hormone levels and blood flow, leading to ED.

Anxiety

Anxiety, particularly performance anxiety, can prevent an erection. Worrying about sexual performance can create a cycle of ongoing ED, as the anxiety itself can cause difficulties.

Depression

Depression can reduce sexual desire and lead to ED. Additionally, many medications used to treat depression have side effects that can cause ED.

Relationship Issues

Interpersonal issues such as lack of communication, unresolved conflicts, and emotional distance from a partner can lead to ED. Relationship counseling may be beneficial in these cases.

Lifestyle Factors

Certain lifestyle choices can increase the risk of developing ED.

Smoking

Smoking can damage blood vessels and restrict blood flow

to the penis. It is a significant risk factor for ED, particularly in men with vascular disease.

Alcohol Consumption

While moderate alcohol consumption may not harm erectile function, excessive drinking can lead to ED. Alcohol can depress the central nervous system and interfere with the signals necessary for an erection.

Drug Use

The use of recreational drugs such as marijuana, cocaine, and methamphetamines can cause ED. These substances can damage blood vessels and reduce blood flow to the penis.

Lack of Physical Activity

A sedentary lifestyle is associated with obesity, cardiovascular disease, and diabetes, all of which are risk factors for ED. Regular physical activity improves overall health and can help prevent ED.

Medications and ED

Certain medications can cause or contribute to ED as a side effect.

Common Medications that Cause ED

- Antihypertensives: Some blood pressure medications, such as beta-blockers and diuretics, can cause ED.

- Antidepressants: Selective serotonin reuptake inhibitors (SSRIs) and other antidepressants can interfere with sexual function.

- Antihistamines: Some allergy medications can cause ED.

- Antipsychotics: Medications used to treat psychiatric disorders can have sexual side effects.

- Chemotherapy: Cancer treatments can affect hormonal balance and nerve function, leading to ED.

Managing Medication-Induced ED

If medication is causing ED, do not stop taking it without consulting a healthcare provider. Management strategies may include:

- Adjusting the Dose: Sometimes, a lower dose of the medication can reduce side effects while still providing therapeutic benefits.

- Switching Medications: Your healthcare provider may suggest an alternative medication that has fewer sexual side effects.

- Additional Treatments: Medications such as phosphodiesterase type 5 (PDE5) inhibitors (e.g., sildenafil, tadalafil) can help manage ED symptoms.

-    Lifestyle   Changes: Improving diet, increasing physical activity, reducing stress, and quitting smoking can help mitigate the effects of medication-induced ED.

Understanding the various causes of Erectile Dysfunction can help in diagnosing and treating this condition effectively.

Chapter 4: Diagnosis of Erectile Dysfunction

Initial Consultation

The initial consultation involves discussing the patient's symptoms, concerns, and overall health. It provides an opportunity for the healthcare provider to gather information, understand the context of the ED, and decide on the appropriate diagnostic approach. Key elements include:

- Patient's Concerns: Discussing the nature, duration, and severity of the erectile difficulties.

- Impact on Life: Exploring how ED affects personal and sexual relationships.

- Previous Treatments: Reviewing any previous attempts at treatment and their outcomes.

Medical History

A comprehensive medical history is crucial for identifying potential physical or systemic causes of ED. It includes:

- Chronic Conditions: Inquiring about conditions such as diabetes, cardiovascular diseases, or kidney disease.

- Medications: Reviewing all current and past medications to identify any that may contribute to ED.

- Surgical History: Considering any surgeries, particularly those involving the pelvic region, prostate, or bladder.

Sexual History

Understanding the patient's sexual history helps pinpoint whether the ED is consistent or situational and if it affects only specific scenarios or partners. Important aspects to cover include:

- Onset and Duration: When ED began and whether it's persistent or intermittent.

- Sexual Function: Details about erections, sexual desire, and performance issues.

- Partner Factors: Whether ED occurs with all partners or specific ones, and any recent changes in relationships.

Physical Examination

A physical examination assesses overall health and identifies any physical causes of ED. Components include:

- Genital Examination: Checking for anatomical abnormalities, Peyronie's disease, or signs of trauma.

- Vascular Examination: Assessing pulses in the groin,

abdomen, and legs to evaluate blood flow.

- Neurological Examination: Testing reflexes and sensation in the pelvic region to assess nerve function.

Diagnostic Tests

Diagnostic tests help determine the underlying causes of ED by evaluating physiological and biochemical factors.

Blood Tests

Blood tests can identify hormonal imbalances,

diabetes, and other systemic conditions affecting erectile function. Common tests include:

- Testosterone Levels: Measuring testosterone to check for low levels that could contribute to ED.

- Blood Glucose Levels: Checking for diabetes or prediabetes.

- Lipid Profile: Assessing cholesterol and triglycerides to evaluate cardiovascular health.

Urinalysis

A urinalysis can reveal underlying conditions such as diabetes or kidney disease. It involves examining urine for signs of glucose, protein, or other abnormalities.

Ultrasound

Ultrasound, particularly Doppler ultrasound, evaluates blood flow in the penis. It can help identify vascular issues such as insufficient blood flow or venous leakage. The procedure involves:

- Penile Doppler Ultrasound: Assessing blood flow before

and after administration of a vasodilator to evaluate the responsiveness of penile blood vessels.

Nocturnal Penile Tumescence (NPT) Test

The NPT test assesses whether erections occur during sleep, which can help distinguish between physical and psychological causes of ED. The test involves:

- Using a Device: A special device with a ring or sensors is placed around the penis to measure the frequency and rigidity of nocturnal erections.

- Interpreting Results: The presence of nocturnal erections typically suggests a physical cause, while their absence might indicate a psychological issue.

Psychological Evaluation

A psychological evaluation can help determine if psychological factors are contributing to ED. It involves:

- Assessing Mental Health: Evaluating for conditions such

as anxiety, depression, and stress that could impact sexual function.

- Exploring Relationship Issues: Discussing any interpersonal or relationship problems that might affect sexual performance.

By combining these diagnostic approaches, healthcare providers can develop a comprehensive understanding of the factors contributing to ED and create an effective treatment plan tailored to the individual's needs.

Chapter 5: Treatment Options for Erectile Dysfunction

Lifestyle Modifications

Lifestyle changes are often the first line of treatment for ED, especially when addressing contributing factors such as poor health habits or stress.

Diet and Nutrition

A healthy diet can improve cardiovascular health, hormone levels, and overall well-being, all of which are crucial for erectile function. Recommendations include:

- Balanced Diet: Emphasizing fruits, vegetables, whole grains, lean proteins, and healthy fats.

- Limiting Processed Foods: Reducing intake of high-fat, high-sugar, and processed foods that can contribute to obesity and cardiovascular disease.

- Hydration: Ensuring adequate fluid intake for overall health.

Exercise

Regular physical activity can enhance cardiovascular health, reduce weight, and improve mood, all of which can help with ED. Recommended activities include:

- Aerobic Exercise: Activities such as walking, jogging, cycling, or swimming to improve cardiovascular function.

- Strength Training: Building muscle mass and improving overall physical fitness.

- Pelvic Floor Exercises: Strengthening pelvic muscles through exercises like Kegels.

Stress Management

Managing stress through various techniques can help alleviate psychological contributors to ED. Strategies include:

- Relaxation Techniques: Practicing deep breathing, meditation, or yoga to reduce stress.

- Time Management: Implementing effective time management and relaxation techniques to alleviate work-related stress.

Oral medications are commonly prescribed to help achieve and maintain an erection by increasing blood flow to the penis.

Phosphodiesterase-5 (PDE5) Inhibitors

PDE5 inhibitors are the most widely used medications for ED. They work by enhancing the effects of nitric oxide, which increases blood flow to the penis. Common PDE5 inhibitors include:

- Sildenafil (Viagra): Typically taken 30 minutes to 1 hour before sexual activity.

- Tadalafil (Cialis): Can be taken daily or on an as-needed basis, with effects lasting up to 36 hours.

- Vardenafil (Levitra): Taken 1 hour before sexual activity.

- Avanafil (Stendra): Works within 15 minutes and has a shorter duration.

Other Medications

In some cases, other medications might be used if PDE5 inhibitors are not effective or suitable. These include:

- Alprostadil: Available as an injection or urethral suppository to enhance blood flow directly.

Hormone Therapy

Hormone therapy is used when ED is linked to hormonal imbalances, such as low testosterone levels. Options include:

- Testosterone Replacement Therapy (TRT): Administered via injections, patches, gels, or oral tablets to normalize testosterone levels.

Intracavernosal Injections

Intracavernosal injections involve injecting medication directly into the penis to induce an erection. Common medications include:

- Alprostadil (Caverject, Edex): A prostaglandin E1 that relaxes blood vessels and increases blood flow to the penis.

- Combination Therapy: Sometimes used to improve efficacy and reduce side effects.

Urethral Suppositories

Urethral suppositories are another method of delivering medication to induce an erection. Alprostadil is commonly used in this form:

- MUSE (Medicated Urethral System for Erection): A small suppository inserted into the urethra that can cause an erection within 10 minutes.

For cases where other treatments are ineffective, mechanical devices or surgical options may be considered.

Vacuum Erection Devices (VED)

VEDs use a vacuum pump to draw blood into the penis and then maintain the erection with a constriction band. They are a non-invasive option and can be effective in achieving an erection.

Penile Implants

Penile implants are surgically placed devices that can provide a permanent solution to ED. Options include:

- Inflatable Implants: Consist of cylinders implanted in the penis that can be inflated with a pump to achieve an erection.

- Malleable Implants: Bendable rods that can be manually adjusted to create an erection.

Vascular Surgery

Vascular surgery may be considered in cases of significant vascular issues. Procedures include:

- Arterial Revascularization: Improving blood flow to the penis by bypassing blocked arteries.

- Venous Ligation: Reducing blood flow out of the penis to maintain an erection.

Psychological Counseling

Psychological factors play a significant role in ED, and addressing these can be crucial for effective treatment.

Cognitive Behavioral Therapy (CBT)

CBT focuses on changing negative thought patterns and behaviors that contribute to ED. It can help manage anxiety, depression, and performance issues.

Sex Therapy

Sex therapy addresses sexual concerns and improves communication between partners. It can help overcome sexual dysfunction and enhance sexual satisfaction.

Relationship Counseling

Relationship counseling can improve communication and resolve conflicts between partners, which may contribute to ED. It focuses on enhancing emotional and sexual intimacy.

Alternative and Complementary Therapies

Some individuals explore alternative or complementary therapies for ED.

Acupuncture

Acupuncture involves inserting fine needles into specific points on the body to improve energy flow and alleviate various conditions, including ED. Some studies suggest it may help with blood flow and sexual function.

Herbal Supplements

Herbal supplements are sometimes used to improve erectile function, although evidence varies in terms of efficacy and safety. Common supplements include:

- Ginseng: May improve sexual function and desire.

- Ginkgo Biloba: Thought to enhance blood flow and sexual function.

- Yohimbine: Derived from the bark of the yohimbe tree, it may have effects on sexual function, though its use is controversial due to potential side effects.

It is essential to consult with a healthcare provider before starting any new treatments or supplements, as some may interact with medications or underlying health conditions.

Chapter 6: Living with Erectile Dysfunction

Impact on Relationships

Erectile Dysfunction (ED) can significantly affect relationships, influencing both partners' emotional well-being and the dynamics of intimacy.

- Emotional Strain: ED can lead to feelings of frustration, inadequacy, and reduced self-esteem. These feelings can be challenging for both the person experiencing ED and their partner.

- Intimacy Issues: The inability to achieve or maintain an erection can affect physical intimacy, which may lead to emotional distance or dissatisfaction in the relationship.

- Communication Challenges: Difficulties in discussing ED can create misunderstandings or feelings of rejection between partners.

Communication with Partners

Effective communication is crucial for managing the

impact of ED on relationships. Strategies include:

- Open Dialogue: Share your feelings and concerns honestly with your partner. Explain how ED affects you and discuss any anxieties or frustrations.

- Reassurance: Reassure your partner that ED is a common condition and not a reflection of their attractiveness or your feelings for them.

- Explore Alternatives: Discuss other ways to express intimacy and maintain a connection, such as through non-sexual physical affection, emotional bonding, or trying new activities together.

Coping Strategies

Coping with ED involves both practical and emotional approaches:

- Educate Yourself: Understanding the condition and available treatments can help reduce anxiety and empower you to make informed decisions about your health.

- Seek Professional Help: Consult a healthcare provider for diagnosis and treatment options. Working with a therapist or counselor can also help address psychological aspects of ED.

- Implement Lifestyle Changes: Adopt healthy lifestyle habits, such as a balanced diet, regular exercise, and stress management techniques, to improve overall health and potentially alleviate ED.

- Maintain a Positive Outlook: Focus on the aspects of your

life that are fulfilling and rewarding. Developing a positive attitude can help manage stress and improve your overall well-being.

Support Groups and Resources

Support groups and resources can provide valuable assistance and reassurance for those living with ED:

- Support Groups: Joining support groups, either in-person or online, can connect you with others experiencing similar challenges. Sharing experiences and advice can provide emotional support and practical tips.

- Educational Resources: Many organizations and websites offer information about ED, including treatment options, coping strategies, and research updates. Reliable sources include:

- American Urological Association (AUA)

- National Institute of Diabetes and Digestive and Kidney Diseases (NIDDK)

- Mayo Clinic and other reputable medical centers

- Professional Counseling: Engage with counselors or therapists who specialize in sexual health or relationship issues. They can offer guidance and strategies to manage the psychological impact of ED.

By addressing ED openly, utilizing coping strategies, and accessing support resources, individuals and couples can better manage the condition and maintain healthy, fulfilling relationships.

Chapter 7: Prevention of Erectile Dysfunction

Preventing Erectile Dysfunction (ED) involves adopting a holistic approach to health that addresses both physical and psychological factors. Here are key strategies for reducing the risk of ED:

Healthy Lifestyle Choices

Maintaining a healthy lifestyle is crucial for preventing ED and promoting overall well-being.

- Balanced Diet: Eat a diet rich in fruits, vegetables, whole grains, lean proteins, and healthy fats. Avoid excessive consumption of processed foods, sugary drinks, and saturated fats.

- Regular Exercise: Engage in regular physical activity, such as aerobic exercises (walking, jogging, cycling) and strength training. Exercise improves cardiovascular health, maintains a healthy weight, and enhances mood.

- Healthy Weight: Achieve and maintain a healthy weight to

reduce the risk of obesity, which is associated with various health conditions that can contribute to ED.

- Limit Alcohol: Consume alcohol in moderation. Excessive drinking can impair sexual function and lead to ED.

- Avoid Smoking: Smoking damages blood vessels and restricts blood flow, increasing the risk of ED. Quitting smoking improves cardiovascular health and reduces the risk of erectile issues.

Regular Medical Checkups

Routine medical checkups are essential for early detection and management of conditions that can lead to ED.

-       Routine Screenings: Regularly monitor blood pressure, cholesterol levels, and blood sugar levels to detect and manage risk factors for cardiovascular disease and diabetes.

- Annual Physical Exams: Schedule annual physical exams to assess overall

health and address any
emerging health issues.

- Discuss Sexual Health: Don't
hesitate to discuss sexual
health with your healthcare
provider. Early intervention
can address potential issues
before they develop into ED.

Managing Chronic Conditions

Effective management of
chronic conditions can help
prevent ED by minimizing
their impact on sexual health.

- Diabetes Management: Maintain stable blood sugar levels through medication, diet, and lifestyle changes to prevent complications that affect erectile function.

- Cardiovascular Health: Manage cardiovascular risk factors, such as hypertension and high cholesterol, through lifestyle changes and medication as prescribed.

- Hormonal Balance: Address hormonal imbalances with appropriate treatments if diagnosed. For example, testosterone replacement therapy may be necessary if low testosterone levels are detected.

Mental Health and Well-being

Mental health plays a significant role in sexual function, and managing psychological factors is crucial for preventing ED.

- Stress Management: Practice stress-reducing techniques such as mindfulness, meditation, and relaxation exercises to manage daily stress.

- Seek Support: If experiencing anxiety, depression, or other emotional issues, seek professional help from a mental health provider. Therapy or counseling can address underlying psychological factors contributing to ED.

- Healthy Relationships: Foster open communication and a strong emotional connection with your partner. Healthy relationships can enhance intimacy and reduce performance anxiety.

By implementing these preventive measures, individuals can reduce their risk of developing ED and maintain overall health and well-being.

The field of Erectile Dysfunction (ED) research is continuously evolving, with ongoing advancements promising to improve diagnosis, treatment, and understanding of the condition. Here's a look at some key areas of focus and future directions:

Emerging Treatments

1. Gene Therapy: Gene therapy aims to address ED

by targeting the underlying genetic and molecular causes of the condition. Research is exploring methods to deliver genes that could enhance nitric oxide production or improve blood vessel function directly to the penile tissues.

2. Stem Cell Therapy: Stem cell therapy is being investigated for its potential to regenerate damaged tissues and improve erectile function. Studies are exploring the use of stem cells to repair or replace damaged

cells in the penile tissues and blood vessels.

3. New Medications: Researchers are developing new oral and injectable medications that may offer improved efficacy and fewer side effects compared to existing treatments. This includes new PDE5 inhibitors with enhanced action profiles and novel compounds targeting different pathways involved in erectile function.

4. Topical Treatments: Research is ongoing into topical treatments, such as

gels or creams, that could be applied directly to the penis to improve blood flow and enhance erectile function without the need for systemic medication.

Advances in Medical Technology

1. Improved Diagnostic Tools: Advances in imaging technology and diagnostic methods are enhancing the ability to diagnose ED more accurately. New techniques in penile Doppler ultrasound, magnetic resonance imaging

(MRI), and other diagnostic tools are providing better insights into the physiological and vascular causes of ED.

2. Wearable Devices: Wearable technology, such as smartwatches or other sensors, is being explored to monitor physiological parameters and track changes in erectile function. These devices may help in early detection and ongoing management of ED.

3. Robotic Surgery: Robotic-assisted surgery is becoming more sophisticated, offering

less invasive options for penile implants and vascular surgeries. Improved precision and reduced recovery times are making these surgical options more accessible.

Ongoing Clinical Trials

Clinical trials are critical for advancing the understanding and treatment of ED. Current and upcoming trials are focusing on:

1. Innovative Therapies: Clinical trials are investigating new drugs, biologics, and devices aimed at treating ED.

These studies evaluate the safety, efficacy, and optimal use of emerging therapies.

2. Combination Therapies: Trials are exploring the effectiveness of combining different treatment modalities, such as combining oral medications with psychological therapy or lifestyle interventions, to achieve better outcomes.

3. Personalized Medicine: Research is increasingly focusing on personalized approaches to ED treatment. Trials are examining how

genetic, hormonal, and individual factors can influence treatment responses and outcomes, aiming to tailor therapies to individual needs.

4. Long-Term Outcomes: Studies are assessing the long-term effects of various ED treatments, including the durability of response, side effects, and overall impact on quality of life.

By staying informed about these advancements and participating in clinical trials, patients and healthcare providers can benefit from the latest research and potentially more effective treatments for ED.

Chapter 9: Frequently Asked Questions (FAQs) about Erectile Dysfunction

Common Concerns about ED

1. What causes Erectile Dysfunction (ED)?

ED can be caused by a variety of factors, including physical conditions like cardiovascular disease, diabetes, and hormonal imbalances, as well as psychological issues such as stress, anxiety, and depression. Lifestyle factors such as smoking, excessive alcohol consumption, and lack of exercise can also contribute.

2. How is ED diagnosed?

Diagnosis typically involves a combination of medical history, physical examination, and diagnostic tests. Tests may include blood tests, urinalysis, ultrasound, nocturnal penile tumescence (NPT) test, and psychological evaluations to determine the underlying cause.

3. What are the treatment options for ED?

Treatment options include lifestyle modifications, oral medications (such as PDE5 inhibitors), hormone therapy, intracavernosal injections, urethral suppositories, vacuum erection devices, penile implants, and psychological counseling. The choice of treatment depends on the underlying cause and individual preferences.

4. Can ED be prevented?

ED can often be prevented by adopting a healthy lifestyle, including a balanced diet, regular exercise, maintaining a healthy weight, avoiding smoking and excessive alcohol, and managing chronic conditions such as diabetes and hypertension.

5. Are there any side effects of ED medications?

Common side effects of PDE5 inhibitors include headaches, flushing, indigestion, and nasal congestion. Other medications may have different side effects. It's important to discuss potential side effects with a healthcare provider to find the most suitable treatment.

Myths and Misconceptions

1. Myth: ED is only a problem for older men.

Fact: While ED is more common in older men, it can affect men of all ages. Young men can experience ED due to psychological factors, lifestyle issues, or underlying health conditions.

2. Myth: ED is always caused by psychological issues.

Fact: While psychological factors can contribute to ED, many cases have physical causes such as cardiovascular disease, diabetes, or hormonal imbalances. Often, ED is the result of a combination of physical and psychological factors.

3. Myth: Only men who have serious health problems experience ED.

Fact: ED can occur in men with no major health issues. Lifestyle factors, temporary stress, or relationship problems can also contribute to ED. It's important to address both physical and non-physical factors.

4. Myth: ED medications are addictive or harmful.

Fact: PDE5 inhibitors are not addictive. They are generally safe when used as prescribed, but it's important to use them under the guidance of a healthcare provider. They should not be taken with certain other medications or by individuals with specific health conditions without medical supervision.

5. Myth: ED means that a man has low libido or isn't interested in sex.

Fact: ED is about the ability to achieve or maintain an erection, not necessarily about sexual desire. Many men with ED still have a normal or high libido but may struggle with achieving an erection.

6. Myth: ED is a normal part of aging and nothing can be done about it.

Fact: While ED becomes more common with age, it is not an

inevitable part of aging. Many effective treatments are available that can help manage or reverse the condition, depending on its cause.

Addressing these common concerns and myths helps to provide clarity and reduce stigma associated with ED, encouraging individuals to seek appropriate help and treatment.

THE END

www.ingramcontent.com/pod-product-compliance
Lightning Source LLC
Chambersburg PA
CBHW071941210526
45479CB00002B/776